HOW EVOLUTION FLUNKED THE SCIENCE TEST

JOE CREWS

LIBRARY OF SERMONS #5

Published by:
Amazing Facts, Inc.
P.O. Box 1058
Roseville, CA 95678-8058
800-538-7275

Cover design by Matthew Moores
Text design by Greg Solie – Altamont Graphics

ISBN 1-58019-046-4

HOW EVOLUTION FLUNKED THE SCIENCE TEST

JOE CREWS

Recently I talked to a man with a fantastic amount of faith. Not one shade of doubt crept into his animated description of man's origin and destiny. He was an evolutionist I met on an airplane. With incredible confidence he bridged the eons of prehistoric time to explain the existence of modern plant and animal life. His detailed description of human ascent from a tiny, one-celled monad was so vivid and convincing that one could almost believe he had seen the microscopic amoeba turn into a man.

What is this evolution doctrine that inspires so much faith in its disciples? How has it turned great scientists into dogmatic opponents of any other viewpoint? Many evolutionary scientists have united their professional influence to forbid any classroom instruction contrary to their own views. Does the theory of evolution merit this kind of fanatical support, which would silence all opposing ideas? When religious people take such a position, they are called bigots, but scientists seem to escape that charge. In February of 1977, nearly 200 members of the nation's academic community sent letters to school boards across the United States, urging that no alternate ideas on origins be permitted in classrooms.

This indicates that the evolutionists are feeling the threat of a rising revolt against the stereotyped, contradictory versions of their theory. Many students are looking for honest answers to their questions about the origin and purpose of life. For the first time, the stale traditions of evolution have to go on the defensive. But let's take a look at what they have to defend. Then you will understand why these evolutionary scientists are people of such extraordinary faith, and why they are so fearful of facing competition at the school level.

1. Spontaneous Generation

How does the evolutionist explain the existence of that first one-celled animal from which all life forms supposedly evolved? For many years the medieval idea of spontaneous generation was the accepted explanation. According to Webster, spontaneous generation is "the generation of living from nonliving matter … [it is taken] from the belief, now abandoned, that organisms found in putrid organic matter arose spontaneously from it."

Simply stated, this means that under the proper conditions of temperature, time, place, etc., decaying matter simply turns into organic life. This simplistic idea dominated scientific thinking until 1846, when Louis Pasteur completely shattered the theory by his experiments. He exposed the whole concept as utter foolishness. Under controlled laboratory conditions, in a semi-vacuum, no organic life ever emerged from decaying, nonliving matter. Reluctantly it was abandoned as a valid scientific issue. Today no reputable scientist tries to defend it on a demonstrable basis. That is why Webster says it is "now abandoned." It never has been and never can be demonstrated in the test tube. No present process is observed that could support the idea of spontaneous generation. Obviously, if spontaneous generation actually did take place in the distant past to produce the first spark of life, it must be assumed that the laws that govern life had to be completely different from what they are now. But wait a minute! This won't work either, because the whole evolutionary theory rests upon the assumption that conditions on the earth have remained uniform throughout the ages.

Do you begin to see the dilemma of the evolutionists in explaining that first amoeba, or monad, or whatever formed the first cell of life? If it sprang up spontaneously from no previous life, it contradicts a basic law of nature that forms the foundation of the entire theory. Yet, without believing in spontaneous generation, the evolutionist would have to acknowledge something other than natural forces at work—in other words, God. How do they get around this dilemma?

Dr. George Wald, Nobel Prize winner of Harvard University, states it as cryptically and honestly as an evolutionist can:

> "One has only to contemplate the magnitude of this task to concede that the spontaneous generation of a living organism is impossible. Yet here we are—as a result, I believe, of spontaneous generation" (*Scientific American,* August 1954).

That statement by Dr. Wald demonstrates a much greater faith than a religious creationist can muster. Notice that the great evolutionary scientist says it could not have happened. It was impossible. Yet he believes it did happen. What can we say to that kind of faith? At least the creationist believes that God was *able* to speak life into existence. His is not a blind faith in something that he concedes to be impossible.

So here we are, face to face with the first contradiction of evolution with a basic law of science. In order to sustain his humanistic explanation of the origin of life, he must accept the exploded, unscientific theory of spontaneous generation. And the big question is this: Why is he so violently opposed to the spontaneous generation spoken of in the Bible? A miracle of creation is required in either case. Either God did it by divine fiat, or blind, unintelligent nature produced

Wald's impossible act. Let any reasonable mind contemplate the alternatives for a moment. Doesn't it take more faith to believe that chance could produce life than it does to believe infinite intelligence could produce it?

Why did Dr. Wald say that it was impossible for life to result from spontaneous generation? That was not an easy concession for a confirmed evolutionist to make. His exhaustive search for a scientific explanation ended in failure, as it has for all other evolutionary scientists, and he had the courage to admit it. But he also had an incredible faith to believe in it even though it was a scientific impossibility. A Christian who confessed to such a faith would be labeled as naïve and gullible. What a difference the cloak of higher education makes upon our easily impressed minds! How much simpler and sweeter the faith that accepts the inspired account: "In the beginning God created the heaven and the earth" (Genesis 1:1).

2. Chance Life—A Ridiculous Improbability

What would be involved in the accidental development of a single living cell? The fact is that the most elementary form of life is more complicated than any man-made thing on earth. The entire complex of New York City is less complicated than the makeup of the simplest microscopic cell. It is more than ridiculous to talk about its chance production. Scientists themselves assure us that the structure of a single cell is unbelievably intricate. The chance for a proper combination of molecules into amino acids, and then into proteins with the properties of life is entirely unrealistic. *American Scientist* magazine made this admission in January of 1955:

> "From the probability standpoint, the ordering of the present environment into a single amino acid molecule would be utterly improbable in all the time and space available for the origin of terrestrial life."

A Swiss mathematician, Charles Eugene Guye, actually computes the odds against such an occurrence at only one chance in 10(160). That means 10 multiplied by itself 160 times, a number too large even to articulate. Another scientist expressed it this way:

> "The amount of matter to be shaken together to produce a single molecule of

> protein would be millions of times greater than that in the whole universe. For it to occur on earth alone would require many, almost endless, billions of years" (*The Evidence of God in an Expanding Universe*, p. 23).

How can we explain the naïve insistence of evolutionists to believe something so extremely out of character for their scientific background? And how can we harmonize the normally broad-minded tolerance of the educated, with the narrow bigotry exhibited by many evolutionary scientists in trying to suppress opposing points of view? The obvious explanation would seem to be rooted in the desperation of such evolutionists to retain their reputation as the sole dispensers of dogmatic truth. To acknowledge a superior wisdom has been too long cultivated by the evolutionist community. They have repeated their assumptions for so long in support of their theories that they have started accepting them as facts. No one objects to their assuming whatever they want to assume, but to assume happenings that go contrary to all scientific evidence and still call it science is being dishonest.

3. Mutations—How Big the Changes?

Now let's look at a second basic evolutionary teaching which is contrary to scientific law. One of the most necessary parts of evolution, which is supposed to provide the power for changing the amoeba into a man, is *mutation.* This refers to abnormal changes in the organism that are assumed to be caused by chemical changes in the genes themselves. The genes are the hereditary factors within the chromosomes of each species. Every species has its own particular number of chromosomes that contain the genes. Within every human being are 46 chromosomes containing an estimated 100,000 genes, each one of which is able to affect in some way the size, color, texture, or quality of the individual. The assumption is that these genes, which provide the inherited characteristics we get from our ancestors, occasionally become affected by unusual pairing, chemical damage, or other influences, causing them to produce an unusual change in one of the offspring. This is referred to as a mutation. Through gradual changes wrought in the various species through mutation, it is assumed by the evolutionists that the amoeba turned into an invertebrate, which became an amphibian, then a reptile, a quadruped, an ape form, and finally a man. In other words, the species are not fixed in the eyes of the evolutionists. Families are forever drifting over into another higher form as time progresses. This means that all the fossil records of animal history should reveal an utter absence of precise family boundaries. Everything should be in the process of changing into something else—with literally hundreds of millions of half-developed fish trying to become amphibious, and reptiles halfway transformed into birds, and mammals looking like half-apes or half-men.

Now everybody knows that instead of finding those billions of confused family fossils, the scientists have found exactly the opposite. Not one single drifting, changing life form has been studied. Everything stays within the well-defined limits of its own basic kind and absolutely refuses to cooperate with the demands of modern evolutionists. Most people would give up and change their theory when faced with such a crushing, deflating blow, but not the evolutionist! He still searches for that elusive missing link which could at least prove that he hasn't been 100 percent wrong.

But let's look at the vehicle that the evolutionists have depended upon to provide the *possibility* of the drastic changes required by their theory. Sir Julian Huxley, a principal spokesman for evolution, said this:

> "Mutation provides the raw material of evolution." Again he said, "Mutation is the ultimate sources of all … heritable variation" (*Evolution in Action,* p. 38).

Professor Ernst Mayr, another leader of the evolutionists, made this statement:

> "Yet it must not be forgotten that mutation is the ultimate source of all genetic variation found in natural populations and the only raw material available for natural selection to work on" (*Animal Species and Evolution,* p. 170).

Please keep this clearly in mind: Evolutionists say that mutation is absolutely essential to provide the inexorable upgrading of species that changed the simpler forms into more complex forms. BUT—the scientific fact is that mutation could NEVER accomplish what evolution demands of it, for several reasons. As all scientists agree, mutations are very rare. Huxley guesses that only about one in a hundred thousand is a mutant. Secondly, when they do occur, they are almost certain to be harmful or deadly to the organism. In other words, the vast majority of such mutations lead toward extinction instead of evolution; they make the organism worse instead of better. Huxley admits: "The great majority of mutant genes are harmful in their effect on the organism" (*Ibid.*, p. 39).

Other scientists, including Darwin himself, conceded that most mutants are recessive and degenerative; therefore, they would actually be eliminated by natural selection rather than effect any significant improvement in the organism. Professor G. G. Simpson, one of the elite spokesmen for evolution, writes about multiple, simultaneous mutations and reports that the mathematical likelihood of getting good evolutionary results would occur only once in 274 billion years! And that would be assuming 100 million individuals reproducing a new generation *every day!* He concludes by saying:

> "Obviously … such a process has played no part whatever in evolution" (*The Major Features of Evolution,* p. 96).

Does this sound sort of confusing to you? They say mutation is necessary to make the changes required by their theory, yet they have to confess that it is scientifically impossible for multiple mutations to make the changes. This is too typical of the puzzling twists and turns made by our evolutionist friends in their efforts to uphold an exploded theory. So the second point of contradiction with true science has been established.

Mutations, of course, do effect minor changes within the basic kinds, but those changes are limited, never producing a new family. They can explain many of the varieties of both plant and animals but can never explain the creation of basic kinds as required by evolution.

4. Fossils Support Creationism

Since we have discovered that the fossil record gives no support to the idea of species gradually changing into other species, let us see if fossil evidence is in harmony with the Bible. Ten times in the book of Genesis we read God's decree concerning the reproduction of His creatures—*"after its kind."* The word "kind" refers to species, or families. Each created family was to produce only its own kind. This forever precludes the drifting, changing process required by organic evolution where one species turns into another.

Take note that God did not say there could be no changes within the family. He did not create all the varieties of dogs, cats, horses, etc., in the very beginning. There was only a male and female of each species, and many changes have since occurred to produce a wide assortment of varieties within the family. But please keep it straight in your mind that cats have always remained cats, dogs are still dogs, and men are still men. Mutation has only been responsible for producing a new variety of the same species, but never originating another new kind. Selective breeding has also brought tremendous improvements such as hornless cattle, white turkeys, and seedless oranges, but all the organisms continue to reproduce exactly as God decreed at Creation—after its kind.

The "common ancestor" that evolution demands has never existed. There is not a "missing link." Man and monkeys are supposed to stem from the same animal ancestry! Even chimpanzees and many monkey groups vary tremendously. Some are smart, others dumb. Some have short tails and some long. Some have no tails at all. Their teeth vary in number. A few have thumbs and others do not. Their genes are different. Their blood is different. Their chromosomes don't jibe. Interestingly enough, apes only breed with apes, chimpanzees with chimpanzees, and monkeys with monkeys.

But when we start comparing humans with monkeys, we get even more impossible differences than those among the simian types. In fact, these differences constitute another unanswerable support for the Bible rule of "after its kind." The fact that some monkeys can be trained to smoke a pipe, ride a scooter, or even hoist a test tube in a laboratory does not prove that scientists are evolved animals, or that monkeys are retarded, developing humans.

It has already been stated that evolutionists expected the fossil record to support their theory of species changes. Their doctrine demanded vast numbers of scaly reptiles transforming their scales into feathers and their front feet into wings. Other reptiles supposedly should be changing into fur-bearing quadrupeds. Did they find those thousands of multi-changing creatures? Not one! No matter what particular strata they sifted through, all the fossils were easily recognized and classified within their own families, just as God decreed. If the evolutionary doctrine were true, the strata would be teeming with hundreds of millions of transition forms with combination features of two or more species. Not only so, but there would have to be millions upon millions of observable living links *right now* in the process of turning into a higher form. Darwin confessed:

> "There are two or three million species on earth. A sufficient field one might think for observation; but it must be said today that in spite of all the evidence of trained observers, not one change of the species to another is on record" (*Life and Letters,* vol. 3, p. 25).

How interesting! Then why insist that it had to be that way? This is one of the marvels of those who cling to a traditional theory.

Even the most ancient fossil forms in the lowest fossil beds have stubbornly retained the same features of their modern counterparts, and it is amusing to listen to the exclamations of surprise by the evolutionists. The creationist is not surprised at all. His Bible told him it would be that way, and he hasn't been forced to puzzle over contradictory evidence.

5. The Mystery of the Empty Strata

Another frustration for the poor evolutionist is the strange case of the empty strata. As one digs deep into the earth, one layer or stratum after another is revealed. Often we can see these layers clearly exposed in the side of a mountain or roadbed cut. Geologists have given names to the succession of strata that pile one on top of another. Descending into Grand Canyon for example, one moves downward past the Mississippi, Devonian, Cambrian, etc., as the scientists have tagged them.

Now here is the perplexity for the evolutionists: The Cambrian is the last stratum of the descending levels that has any fossils in it. All the lower strata below the Cambrian have absolutely no fossil record of life other than some single-celled types such as bacteria and algae. Why not? The Cambrian layer is full of all the major kinds of animals found today except the vertebrates. In other words, there is nothing primitive about the structure of these most ancient fossils known to man. Essentially, they compare with the complexity of current living creatures. But the big question is: *Where are their ancestors?* Where are all the evolving creatures that should have led up to these highly developed fossils? According to the theory of evolution, the Precambrian strata should be filled with more primitive forms of these Cambrian fossils in the process of evolving upward.

Darwin confessed in his book, *Origin of the Species:*

> "To the question why we do not find rich fossiliferous deposits belonging to these assumed earliest periods prior to the Cambrian system I can give no satisfactory answer … the case at present must remain inexplicable; and may be truly urged as a valid argument against the views here entertained" (p. 309).

How amazing! Darwin admitted having no way to defend his theory, but he still would not adjust his theory to meet the unanswerable arguments against it.

Many other evolutionary scientists have expressed similar disappointment and frustration. Dr. Daniel Axeliod of the University of California calls it:

> "One of the major unsolved problems of geology and evolution" (*Science*, July 4, 1958).

Dr. Austin Clark of the U.S. National Museum wrote concerning the Cambrian fossils:

> "Strange as it may seem … mollusks were mollusks just as unmistakably as they are now" (*The New Evolution: Zoogenesis*, p. 101).

Drs. Marshall Kay and Edwin Colbert of Columbia University marveled over the problem in these words:

> "Why should such complex organic forms be in rocks about 600 million years old and be absent or unrecognized in the records of the preceding two billion years? … If there has been evolution of life, the absence of the requisite fossils in the rocks older than Cambrian is puzzling" (*Stratigraphy and Life History*, p. 102).

George Gaylord Simpson, the "Crown Prince of Evolution," summarized it:

> "The sudden appearance of life is not only the most puzzling feature of the whole fossil record but also its greatest apparent inadequacy" (*The Evolution of Life*, p. 144).

In the face of these forced admissions of failure to find supporting scientific evidence, how can these men of science continue to press so dogmatically for their shaky views? No wonder they fight to keep students from hearing the opposing arguments. Their positions would crumble under the impartial investigation of honest research.

The absence of Precambrian fossils points to one great fact, unacceptable to the evolutionists—a sudden creative act of God that brought all the major creatures into existence at the same time. Their claims that creationism is unscientific are made only to camouflage their own lack of true evidence. The preponderance of physical scientific data is on the side of creation, not evolution.

6. Uniformity or the Flood?

The subject of strata beds leads into the interesting question of how these layers were formed, and why the evolutionists have guesstimated their age in the billions of years. The dating of those layers has been done on the basis of the theory of uniformity. This theory assumes that all the natural processes at work in the past have operated exactly as they do today. In other words, the creation of those strata can only be explained on the basis of what we see happening in the world now. Scientists must calculate how long it takes for sedimentation to build a foot-deep stratum. Then that age is assigned to any 12-inch layer, no matter how deeply located within the earth.

Is that a valid assumption to make? Have all the natural forces of the past been just what we can demonstrate and understand today? How naïve and conceited to compel ages past to conform to our limited observation and experience! We can assume what we please, but it proves absolutely nothing except our own gullibility. The Bible explains very graphically about a Flood that ravaged the face of this earth, covering the highest mountains and completely destroying all plant and animal life outside the ark. The destructive action of the Deluge is expressed by these words in the Bible:

> "The same day were all the fountains of the great deep broken up, and the

windows of heaven were opened. And the rain was upon the earth forty days and forty nights" (Genesis 7:11, 12).

The existence of those strata can be scientifically accounted for in perfect harmony with the Bible record. The universal Flood of Genesis provides a much more reasonable explanation of the strata than evolution's speculations. As the waters receded from the earth, powerful tides and currents carved out the great canyons in a short time. Layers of debris, according to the specific weight, were laid down, compressing plant and animal life into a compact seam or stratum. Only thus can we explain the vast oil reserves and coal beds around the world. These are the result of vegetation and animal bodies being buried under extreme heat and pressure. No such process of fossilization is taking place today. No oil or coal is forming by present natural forces at work. Uniformity fails here.

The fact is, there had to be a gigantic cataclysmic overturn of nature, killing and burying millions of tons of plant and animal life. The position of some fossils standing upright through one or more strata indicates that the process was not slow or age long. The material had to be deposited quickly around the body of the animal, or it could not have remained in its erect position. The flood buried millions of fish, many of them contorted as though suddenly overtaken by a phenomenal force. Marine fossils have been recovered from the highest mountain ranges, and a checklist on other scientific evidences points to a universal deluge over the entire planet.

7. Survival of the Fittest

"Natural selection" is a coined phrase of the evolutionist to describe the survival of the fittest. Simply stated, it is the natural process that enables the strongest of each generation to survive and the weaker, more poorly adjusted ones to die out. The assumption of evolution is that since only the strongest survive to father the next generation, the species will gradually improve, even advancing into other more highly developed states on the evolutionary scale.

Darwin believed that natural selection was the most important factor in the development of his theory. Many of the top teachers of evolution today are hopelessly at odds on the question of how vital it is. Sir Julian Huxley believes in it, as this statement indicates:

> "So far as we know ... natural selection ... is the only effective agency of evolution" (*Evolution in Action,* p. 36).

He is disputed on this by another one of the heavyweights in the field, Dr. Ernst Mayr:

> "Natural selection is no longer regarded as an all-or-none process but rather as a purely statistical concept" (*Animal Species,* p. 7).

G. G. Simpson, who is regarded as the leading interpreter of the theory today, rejects these opposite views. He said,

> "Search for the cause of evolution has been abandoned. It is now clear that evolution has no single cause" (*The Geography of Evolution,* p. 17).

By the way, when you read about the great unity and agreement that exists among the scientists regarding evolution, don't believe a word of it. Each one is busily experimenting with new speculative possibilities as to how the changes took place and then abandoning them as they appear more and more ridiculous. The one basic tenet they do agree on is that there was no divine fiat creation as described in the Bible.

But come back a moment to the matter of natural selection. What is the evidence that it can actually reproduce all the changes involved in the transition from amoeba to man? Is there scientific proof that it can even make one small change? When it comes right down to answering those questions, the spokesmen for evolution do some of the fanciest footwork in semantics you ever saw and make some of the most amazing admissions. Even though Simpson supports natural selection as a factor, he recognizes the scarcity of evidence in these words:

> "It might be argued that the theory is quite unsubstantiated and has status only as a speculation" (*Major Features,* pp. 118, 119).

But listen to Huxley's circular reasoning on it. He says:

> "On the basis of our present know ledge natural selection is bound to produce genetic adaptations: and genetic

> adaptations are thus presumptive evidence for the efficiency of natural selection" (*Evolution in Action,* p. 48).

Did you follow that gem of logic? His proof for natural selection is adaptation or change in the organism, but the change is produced by natural selection! In other words: A=B; therefore B=A. His "proof" proves nothing. Were the changes produced by natural selection, or did he invent natural selection to explain the changes? It is just as likely that the changes produced the natural selection theory. The ludicrous thing is that even the changes from species to species have never been verified. As we have shown already, there is not one shred of fossil evidence or living evidence that any species has changed into another. So Huxley's proof for natural selection are changes which never happened, and the changes which never happened are offered as proof for natural selection. Surely this is the most vacuous logic to be found in a science textbook.

But let us continue with Sir Julian's explanation about the reliability of this natural selection process:

"To sum up, natural selection converts randomness into direction and blind chance into apparent purpose. It operates with the aid of time to produce improvements in the machinery of living, and in the process generates results of a more than astronomical improbability which could have been achieved in no other way" (*Evolution in Action,* pp. 54, 55).

Don't miss the force of that last sentence. The evolutionary changes wrought by natural selection are "astronomically improbable," but because our friend Huxley sees no other way for it to be done, he believes in the astronomically improbable. Poor man! He is wrong when he said the complex order of life today could have been achieved in no other way. God created the wonders of cell and gene and all the millions of processes that leave the Nobel Prize winners baffled.

But since Sir Julian doesn't believe in a divine creation, he has to invent a miracle-working process to explain the existence of these complex creatures—obviously got here somehow. To illustrate the omnipotence of his "natural selection" god, Huxley computed the odds against such a process.

The computations were done on the likelihood of every favorable evolutionary factor being able to produce a horse. Now keep in mind that this is all a chance development through the operation of nature, time, mutation, and natural selection. In his book, *Evolution in Action,* Huxley gave the odds this way:

> "The figure 1 with three million naughts after it: and that would take three large volumes of about 500 pages each, just to print! ... No one would bet on anything so improbable happening; and yet it has happened" (p. 46).

We commented before about the faith of evolutionists to believe in the impossible. Since this figure of compound probability is effectively zero, how can a scientific mind, in the absence of any demonstrable evidence, be so dogmatic in defending his theory? Why did Huxley employ a mathematical formula to illustrate the impossibility of his theory working? Perhaps he used the figures to accent his personal testimony. Just as born-again Christians seek occasions to bear their personal testimony of faith in Christ, Huxley demolishes the scientific possibilities of his theory in order to magnify the personal faith aspect of his personal testimony for the god evolution.

Marshall and Sandra Hall in their book *The Truth—God or Evolution?* share their reaction to Huxley's absurd faith in the chance production of a horse. It will provide a fitting climax of proof that evolution indeed flunked the science test.

> "And, let us remind you who find such odds ridiculous (even if you are reassured by Mr. Huxley), that this figure was calculated for the evolution of a horse! How many more volumes of zeros would be required by Mr. Huxley to produce a human being? And then you would have just one horse and one human being and, unless the mathematician wishes to add in the probability for the evolution of all the plants and animals that are necessary to support a horse and a man, you would have a sterile world where neither could have survived any stage of its supposed evolution! What have we now—the figure 1 followed by a thousand volumes of zeros? Then add another thousand volumes for the improbability of the earth

having all the necessary properties for life built into it. And add another thousand volumes for the improbability of the sun, and the moon, and the stars. Add other thousands for the evolution of all the thoughts that man can have, all the objective and subjective reality that ebbs and flows in us like part of the pulse beat of an inscrutable cosmos!

"Add them all in and you long ago stopped talking about rational thought, much less scientific evidence. Yet, Simpson, Huxley, Dobzhansky, Mayr, and dozens of others continue to tell us that is the way it had to be! They have retreated from all the points which ever lent any semblance of credibility to the evolutionary theory. Now they busy themselves with esoteric mathematical formulations based on population genetics, random drift, isolation, and other ploys which have a probability of accounting for life on earth of minus zero! They clutter our libraries, and press on the minds of people everywhere an animated waxen image of a theory that has been dead for over a decade.

> "Evolution has no claim whatsoever to being a science.
>
> "It is time all this nonsense ceased. It is time to bury the corpse. It is time to shift the books to the humorous fiction section of the libraries" (pp. 39, 40).

These examples of evolutionary folly are only the tip of an iceberg, but they reassure us that we have no cause to be embarrassed for our creationist faith. Millions of Christians have been intimidated by the high-sounding technical language of educated evolutionists, many of whom are vitriolic in their attacks on special creation. What we do need is more information on exposing the loopholes in the evolutionary theory; its base is so riddled with unscientific inconsistencies, often concealed under the gobbledygook of scientific jargon.

To follow our ancestry back through the sons of Adam, "who was the son of God," is so much more satisfying than to search through dismal swamps for bleeping monad forebears. The human race has dropped, even in our lifetime, several degrees deeper into moral perversion and violent disorder. Humanists cite our animal ancestry as an excuse for much of this bizarre behavior. Why blame people for action dictated by their bestial genes and chromosomes? This rationalization, like a temporary insanity plea, provides license for further irresponsible conduct. The true cause for evil and the true remedy for it are found only in the Word of God. Sin has defaced the image of God in man, and only a personal encounter with the perfect Saviour will bring a reversal of the problem of evil.

Library of Sermons

1. Armageddon
2. Can a Saved Man Choose to Be Lost?
3. Does God's Grace Blot Out the Law?
4. Hidden Eyes and Closed Ears
5. How Evolution Flunked the Science Test
6. Is It Easier to Be Saved or to Be Lost?
7. Man's Flicker or God's Flame
8. Satan in Chains
9. Satan's Confusing Counterfeits
10. Spirits From Other Worlds
11. Thieves in the Church
12. Why God Said Remember
13. Why the Old Covenant Failed
14. The High Cost of the Cross
15. Hell-Fire
16. Is It Possible to Live Without Sinning?
17. Blood Behind the Veil
18. Spirits of the Dead
19. The Brook Dried Up
20. Death in the Kitchen
21. The Search for the True Church
22. Is It a Sin to Be Tempted?
23. Is Sunday Really Sacred?
24. Heaven ... Is It for Real?
25. Rendezvous in Space
26. Christ's Human Nature
27. Point of No Return
28. The Rich Man and Lazarus

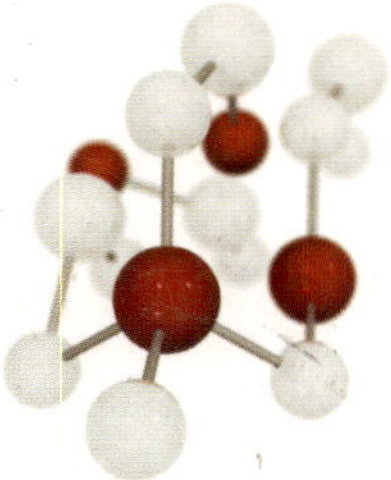

AMAZING FACTS

P.O. Box 1058 • Roseville, CA 95678-

ISBN 1-58019-046-4

9 781580 190466